Learn

Eureka Math™
Grade K
Modules 5 & 6

Published by Great Minds®.

Copyright © 2018 Great Minds®.

Printed in the U.S.A.
This book may be purchased from the publisher at eureka-math.org.
10 9 8 7 6 5 4 3

ISBN 978-1-64054-079-8

GK-M5-M6-L-05.2018

Learn ♦ Practice ♦ Succeed

Eureka Math™ student materials for *A Story of Units*® (K–5) are available in the *Learn, Practice, Succeed* trio. This series supports differentiation and remediation while keeping student materials organized and accessible. Educators will find that the *Learn, Practice,* and *Succeed* series also offers coherent—and therefore, more effective—resources for Response to Intervention (RTI), extra practice, and summer learning.

Learn

Eureka Math Learn serves as a student's in-class companion where they show their thinking, share what they know, and watch their knowledge build every day. *Learn* assembles the daily classwork—Application Problems, Exit Tickets, Problem Sets, templates—in an easily stored and navigated volume.

Practice

Each *Eureka Math* lesson begins with a series of energetic, joyous fluency activities, including those found in *Eureka Math Practice*. Students who are fluent in their math facts can master more material more deeply. With *Practice*, students build competence in newly acquired skills and reinforce previous learning in preparation for the next lesson.

Together, *Learn* and *Practice* provide all the print materials students will use for their core math instruction.

Succeed

Eureka Math Succeed enables students to work individually toward mastery. These additional problem sets align lesson by lesson with classroom instruction, making them ideal for use as homework or extra practice. Each problem set is accompanied by a Homework Helper, a set of worked examples that illustrate how to solve similar problems.

Teachers and tutors can use *Succeed* books from prior grade levels as curriculum-consistent tools for filling gaps in foundational knowledge. Students will thrive and progress more quickly as familiar models facilitate connections to their current grade-level content.

Students, families, and educators:

Thank you for being part of the *Eureka Math*™ community, where we celebrate the joy, wonder, and thrill of mathematics.

In the *Eureka Math* classroom, new learning is activated through rich experiences and dialogue. The *Learn* book puts in each student's hands the prompts and problem sequences they need to express and consolidate their learning in class.

What is in the Learn *book?*

Application Problems: Problem solving in a real-world context is a daily part of *Eureka Math*. Students build confidence and perseverance as they apply their knowledge in new and varied situations. The curriculum encourages students to use the RDW process—Read the problem, Draw to make sense of the problem, and Write an equation and a solution. Teachers facilitate as students share their work and explain their solution strategies to one another.

Problem Sets: A carefully sequenced Problem Set provides an in-class opportunity for independent work, with multiple entry points for differentiation. Teachers can use the Preparation and Customization process to select "Must Do" problems for each student. Some students will complete more problems than others; what is important is that all students have a 10-minute period to immediately exercise what they've learned, with light support from their teacher.

Students bring the Problem Set with them to the culminating point of each lesson: the Student Debrief. Here, students reflect with their peers and their teacher, articulating and consolidating what they wondered, noticed, and learned that day.

Exit Tickets: Students show their teacher what they know through their work on the daily Exit Ticket. This check for understanding provides the teacher with valuable real-time evidence of the efficacy of that day's instruction, giving critical insight into where to focus next.

Templates: From time to time, the Application Problem, Problem Set, or other classroom activity requires that students have their own copy of a picture, reusable model, or data set. Each of these templates is provided with the first lesson that requires it.

Where can I learn more about Eureka Math *resources?*

The Great Minds® team is committed to supporting students, families, and educators with an ever-growing library of resources, available at eureka-math.org. The website also offers inspiring stories of success in the *Eureka Math* community. Share your insights and accomplishments with fellow users by becoming a *Eureka Math* Champion.

Best wishes for a year filled with aha moments!

Jill Diniz

Jill Diniz
Director of Mathematics
Great Minds

The Read–Draw–Write Process

The *Eureka Math* curriculum supports students as they problem-solve by using a simple, repeatable process introduced by the teacher. The Read–Draw–Write (RDW) process calls for students to

1. Read the problem.

2. Draw and label.

3. Write an equation.

4. Write a word sentence (statement).

Educators are encouraged to scaffold the process by interjecting questions such as

- What do you see?

- Can you draw something?

- What conclusions can you make from your drawing?

The more students participate in reasoning through problems with this systematic, open approach, the more they internalize the thought process and apply it instinctively for years to come.

Contents

Module 5: Numbers 10–20 and Counting to 100

Module 6: Analyzing, Comparing, and Composing Shapes

Grade K
Module 5

Marta loves to share her peanuts.

She counted 10 peanuts into the hands of her friend Joey.

Draw a picture of the peanuts in Joey's hands.

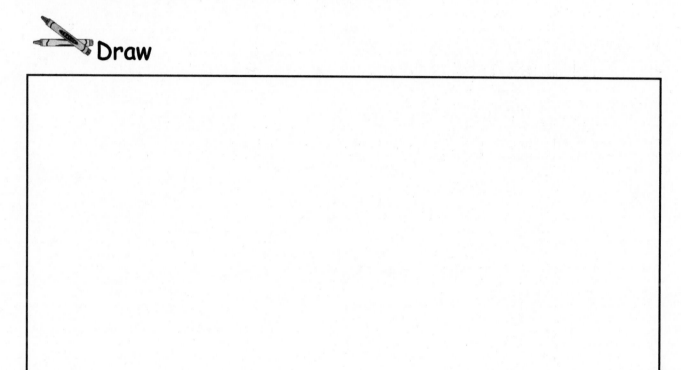Draw

Lesson 1:   Count straws into piles of ten; count the piles as 10 ones.

©2018 Great Minds®. eureka-math.org

3

Name _____   Date _____

Circle the groups that have 10 ones.

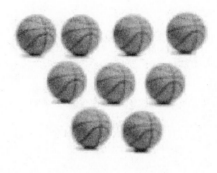

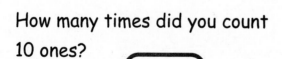

How many times did you count 10 ones?

EUREKA
MATH™

Lesson 1:   Count straws into piles of ten; count the piles as 10 ones.

©2018 Great Minds®. eureka-math.org

5

Name _____   Date _____

Circle the groups that have 10 things.

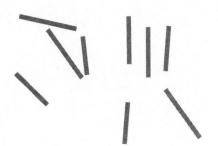

How many times did you count 10 things?

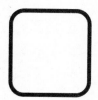

EUREKA
MATH™

Lesson 1:   Count straws into piles of ten; count the piles as 10 ones.

©2018 Great Minds®. eureka-math.org

7

Lisa counted some sticks into one pile of 10.

She counted 5 other sticks into another pile.

Draw a picture to show Lisa's piles of sticks.

Draw

Extension: Have early finishers draw Lisa's piles on another day when she made one pile of 10 sticks and one pile of 8 sticks.

Lesson 2: Count 10 objects within counts of 10 to 20 objects, and describe as 10 ones and __ ones.

©2018 Great Minds®. eureka-math.org

Name _____   Date _____

I have 10 ones and 2 ones.

Touch and count 10 things. Put a check over each one as you count 10 things.

I have 10 ones and ____ ones.

I have 10 ones and ____ ones.

I have ____ ones and ____ ones.

I have ____ ones and ____ ones.

EUREKA MATH™   Lesson 2:   Count 10 objects within counts of 10 to 20 objects, and describe as 10   11
                                  ones and ___ ones.

©2018 Great Minds®. eureka-math.org

Draw pictures to match the words.

I have 10 small circles and 2 small circles:

I have 10 ones and 4 ones:

Lesson 2:   Count 10 objects within counts of 10 to 20 objects, and describe as 10 ones and __ ones.

©2018 Great Minds®. eureka-math.org

EUREKA MATH™

Name _____ Date _____

✬ ✬ ✬ ✬ ✬   10 ones and 3 ones
✬ ✬ ✬ ✬ ✬ ✬   (10 ones and 1 one)

Circle the correct numbers that describe the pictures.

🍎🍎🍎🍎🍎 🍎🍎🍎🍎🍎 🍎🍎🍎	10 ones and 3 ones 10 ones and 7 ones
🍦🍦🍦🍦🍦 🍦🍦🍦🍦 🍦🍦 🍦🍦🍦	10 ones and 8 ones 10 ones and 5 ones
🥨🥨🥨🥨🥨🥨🥨🥨🥨🥨 🥨🥨🥨🥨🥨🥨🥨🥨🥨🥨	10 ones and 10 ones 10 ones and 8 ones
⚪⚪ ⚪⚪ ⚪⚪ ⚪⚪ ⚪⚪ ⚪⚪	10 ones and 4 ones 10 ones and 2 ones

EUREKA
MATH™

Lesson 2: Count 10 objects within counts of 10 to 20 objects, and describe as 10 ones and __ ones.

©2018 Great Minds®. eureka-math.org

13

A gingerbread man has 10 sprinkles as buttons and 2 sprinkles as eyes.
Draw to show the 12 sprinkles as 10 buttons and 2 eyes.

Draw

EUREKA
MATH™

Lesson 3:   Count and circle 10 objects within images of 10 to 20 objects, and
              describe as 10 ones and ___ ones.

©2018 Great Minds®. eureka-math.org

15

Name _____   Date _____

I have 10 ones and 2 ones.

Count and circle 10 things. Tell how many there are in two parts, 10 ones and some more ones.

I have 10 ones and ____ ones.

I have ____ ones and ____ ones.

I have ____ ones and ____ ones.

I have ____ ones and ____ ones.

EUREKA
MATH™

Lesson 3:   Count and circle 10 objects within images of 10 to 20 objects, and describe as 10 ones and __ ones.

©2018 Great Minds®. eureka-math.org

17

Draw your picture to match the words. Circle 10 ones.

I have 10 ones and 3 ones:

I have 10 ones and 8 ones:

Lesson 3:   Count and circle 10 objects within images of 10 to 20 objects, and describe as 10 ones and __ ones.

EUREKA MATH

Name _____   Date _____

Circle 10 ones.   Draw 10 ones and 6 ones.

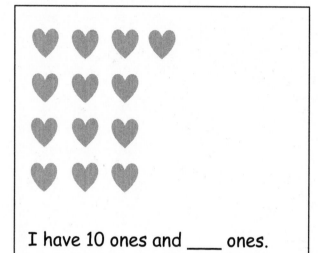

I have 10 ones and ____ ones.   I have 10 ones and 6 ones.

Lesson 3:   Count and circle 10 objects within images of 10 to 20 objects, and
               describe as 10 ones and __ ones.

©2018 Great Minds®. eureka-math.org

19

At recess, 17 students were playing.

10 students played handball while 7 students played tetherball.

 Draw

Draw to show the 17 students as 10 students playing handball and 7 students playing tetherball.

EUREKA MATH™

Lesson 4:   Count straws the Say Ten way to 19; make a pile for each ten.

21

Name _____

Date _____

Draw 10 ones and some ones. Whisper count as you work the Say Ten Way.

I can make ten three.
10   3

I can make ten seven.
10   7

Lesson 4:   Count straws the Say Ten way to 19; make a pile for each ten.

23

EUREKA MATH™

©2018 Great Minds®. eureka-math.org

I can make ten two.
10 2

I can make ten nine.
10 9

Lesson 4:   Count straws the Say Ten way to 19; make a pile for each ten.

EUREKA MATH™

Name _____   Date _____

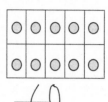

<u>10</u>   <u>3</u>

Count and write how many the Say Ten way.

   ☆

<u>   10   </u>   <u>   </u>   <u>   10   </u>   <u>   </u>

<u>   </u>   <u>   </u>   <u>   </u>   <u>   </u>

EUREKA MATH™

Lesson 4:   Count straws the Say Ten way to 19; make a pile for each ten.

©2018 Great Minds®. eureka-math.org

25

Pat covered 16 holes when playing the flute.

She covered 10 holes with her fingers on the first note she played. She covered 6 holes on the next note she played.

Draw

Draw the 10 holes. Draw the 6 holes. Use your drawing to count all the holes the Say Ten way.

Name _____   Date _____

Ten   two
10   2

Circle 10 things. Touch and count the Say Ten way. Count your 10 ones first. Put a check over the loose ones. Draw a line to match the number.

Ten   one
10   1

Ten   seven
10   7

Ten   three
10   3

Ten   four
10   4

Two   ten
10   10

Ten   eight
10   8

EUREKA MATH™

©2018 Great Minds®. eureka-math.org

Name _____ Date _____

Write and whisper the missing numbers.

Count the Say Ten way from 11 to 20.

<u>10</u> and <u>1</u>	<u>10</u> and <u>2</u>	<u>10</u> and ___	<u>10</u> and <u>4</u>	<u>10</u> and ___
<u>10</u> and <u>6</u>	___ and ___	___ and ___	___ and ___	<u>10</u> and <u>10</u>

There are 18 students: 10 girls and 8 boys.

Draw the 18 students as 10 girls and 8 boys.

Draw

(Note: Remember that the focus is on counting all to find the total rather than counting on or addition.)

Lesson 6:   Model with objects and represent numbers 10 to 20 with place value or Hide Zero cards.

©2018 Great Minds®. eureka-math.org

33

Name _____   Date _____

Write and draw the number. Use your Hide Zero cards to help you.

1 0   3

↓   ↓

1 3

1 0   5

↓   ↓

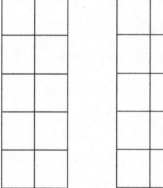

1 0   8

↓   ↓

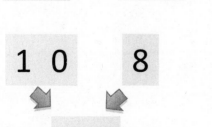

1 0   6

↓   ↓

EUREKA MATH

Lesson 6:   Model with objects and represent numbers 10 to 20 with place value
or Hide Zero cards.

35

©2018 Great Minds®. eureka-math.org

Name _____ Date _____

Draw the number shown on the Hide Zero cards with a drawing in the ten-frame. Write the number below after the 0 is hidden.

Show the number again on the right with a count of 10 ones and 4 ones. Circle the 10 ones.

1 0   **4**

EUREKA MATH™

Lesson 6: Model with objects and represent numbers 10 to 20 with place value or Hide Zero cards.

©2018 Great Minds®. eureka-math.org

37

Gregory drew 10 smiley faces and 5 smiley faces.

He put them together and had 15 smiley faces.

Draw the 15 smiley faces as 10 smiley faces and 5 smiley faces.

Draw

(Give students one Hide Zero card and 5-group cards 1–9.) Now, draw 15 with Hide Zero cards when the zero is hiding and when the zero is not hiding.

EUREKA
MATH™

Lesson 7:   Model and write numbers 10 to 20 as number bonds.

39

©2018 Great Minds®. eureka-math.org

Name _____   Date _____

Look at the Hide Zero cards or the 5-group cards. Use your cards to show the number. Write the number as a number bond.

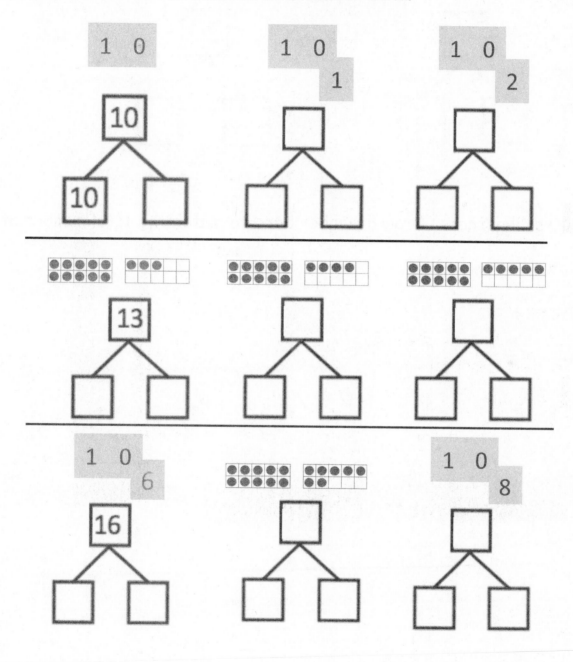

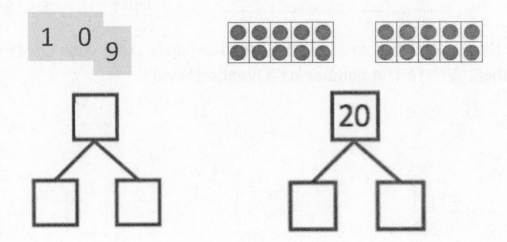

Circle 10 smiley faces. Draw a number bond to match the total number of faces.

EUREKA
MATH™

Name _____ Date _____

Look at the Hide Zero cards or the 5-group cards. Use your cards to show the number. Write the number as a number bond.

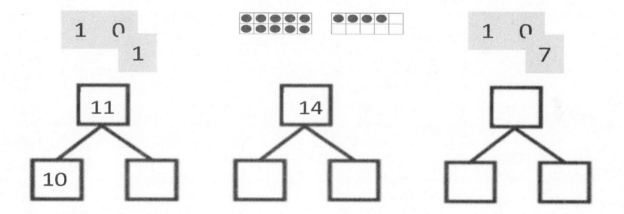

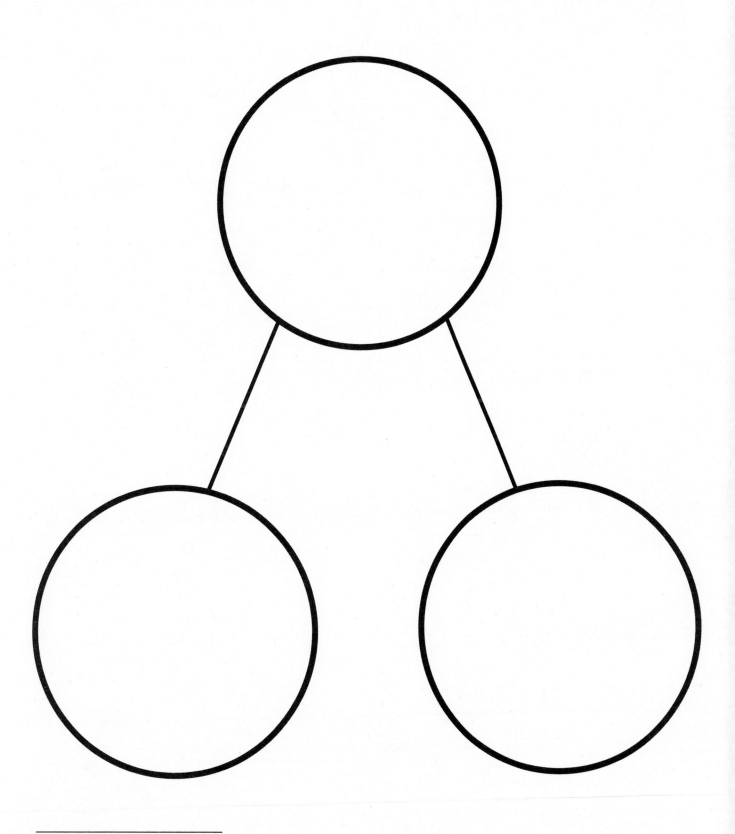

number bond

EUREKA
MATH™

Lesson 7: Model and write numbers 10 to 20 as number bonds.

45

©2018 Great Minds®. eureka-math.org

Peter drew a number bond of 13 as 10 and 3.

Bill drew a number bond, too, but he switched around the 10 and 3.

Show both Bill's and Peter's number bonds.

Draw thirteen things as 10 ones and 3 ones.

Draw

Explain to your partner what you notice about the two number bonds.

Lesson 8:   Model teen numbers with materials from abstract to concrete.

47

Name _____   Date _____

Use your materials to show each number as 10 ones and some more ones. Use your 5-groups way of drawing. Show each number with your Hide Zero cards. Whisper count as you work.

11

18

15

14

12

17

20

13

Lesson 8:   Model teen numbers with materials from abstract to concrete.

EUREKA
MATH™

Name _____   Date _____

Use your materials to show the number as 10 ones and some more ones.
Use your 5-groups way of drawing.

1   6

Use your cubes to show the number. Then, color in the cubes to match the number.

1   2

EUREKA MATH

Lesson 8:   Model teen numbers with materials from abstract to concrete.

©2018 Great Minds®. eureka-math.org

51

Jenny drew 15 things with 1 chip and 5 more chips.

Draw 15 things as 10 ones and 5 ones.

Draw

Explain to your partner why you think Jenny made her mistake.

Name _____   Date _____

Whisper count as you draw the number. Fill one 10-frame first. Show your numbers with your Hide Zero cards.

12

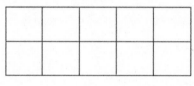

17

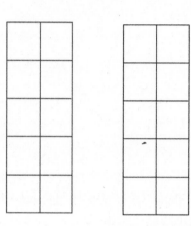

16

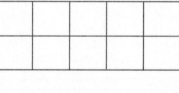
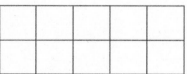

13

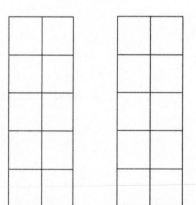

Draw and circle 10 ones and some more ones to show each number.

| 20 | 11 |

Choose a teen number to draw. Circle 10 ones and some ones to show each number.

Lesson 9:   Draw teen numbers from abstract to pictorial.

Name _____   Date _____

Show the number by filling in the 10-frames with circles.

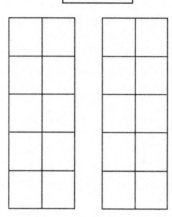

15

19

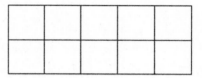

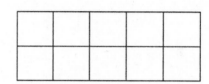

Draw circles to show the number. Circle 10 ones.

18

14

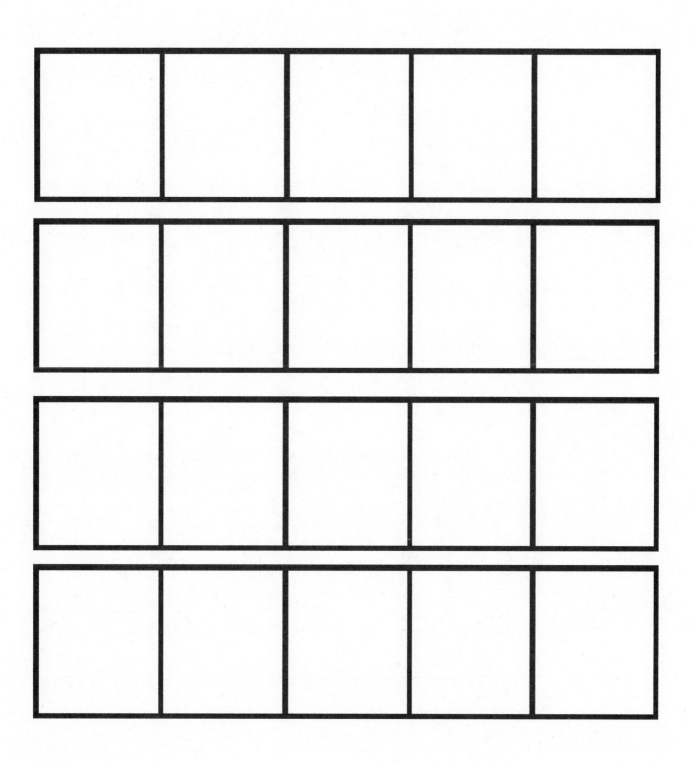

double 10-frame

Ms. Garcia is painting her fingernails. She painted all the nails on her left hand except her thumb. How many more nails does she need to paint? How many does she have left to paint after she paints her left thumb? Draw a picture to help you.

Draw

EUREKA
MATH™

Name _____   Date _____

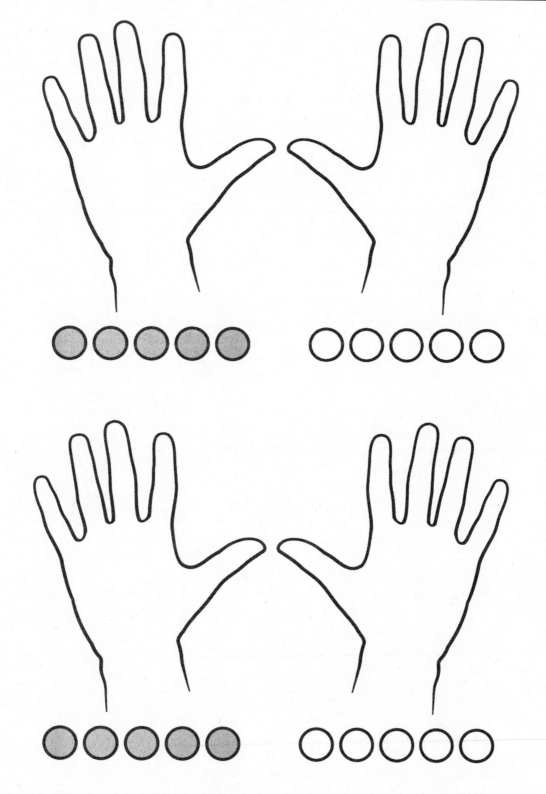

EUREKA
MATH™

Name _____   Date _____

Use your red crayon and yellow crayon to draw the beads from your Rekenrek in two lines.

How many beads did you draw?

Trace your hands. Draw your fingernails. How many fingernails do you have on your two hands?

Mary has 10 toy trucks. She told her mom she likes to spread them out on the floor. She said she doesn't like to put them away neatly in the little toy box because there are fewer toys.

Draw

Draw a picture to prove to Mary that the number of toy trucks is the same when they are all spread out as when they are in the little toy box.

EUREKA MATH™

Lesson 11:   Show, count, and write numbers 11 to 20 in tower configurations increasing by 1—a pattern of 1 larger.

67

©2018 Great Minds®. eureka-math.org

Name _____   Date _____

Count, color and write.

| 10 | 11 | | | 14 | | | | | 19 | |

Lesson 11:   Show, count, and write numbers 11 to 20 in tower configurations
increasing by 1—a pattern of *1 larger*.

©2018 Great Minds®. eureka-math.org

69

Name _____   Date _____

Start at the bottom. Draw lines to put the numbers in order on the tower.
Then, write the numbers in the tower. Say each number the regular way
and the Say Ten way as you work.

12 ●

19 ●

16 ●

14 ●

17 ●

20
18
15
13
11
10

Lesson 11:   Show, count, and write numbers 11 to 20 in tower configurations
increasing by 1—a pattern of *1 larger*.

71

EUREKA
MATH™

©2018 Great Minds®. eureka-math.org

Peter was sitting at lunch eating his french fries. He counted 8 left on his plate. He ate 1 french fry. He ate another french fry. Then, he ate another french fry. How many french fries did Peter have then?

Draw

Lesson 12:   Represent numbers 20 to 11 in tower configurations decreasing by
1—a pattern of 1 smaller.

©2018 Great Minds®. eureka-math.org

73

EUREKA
MATH™

Name _____   Date _____

Count, color and write.

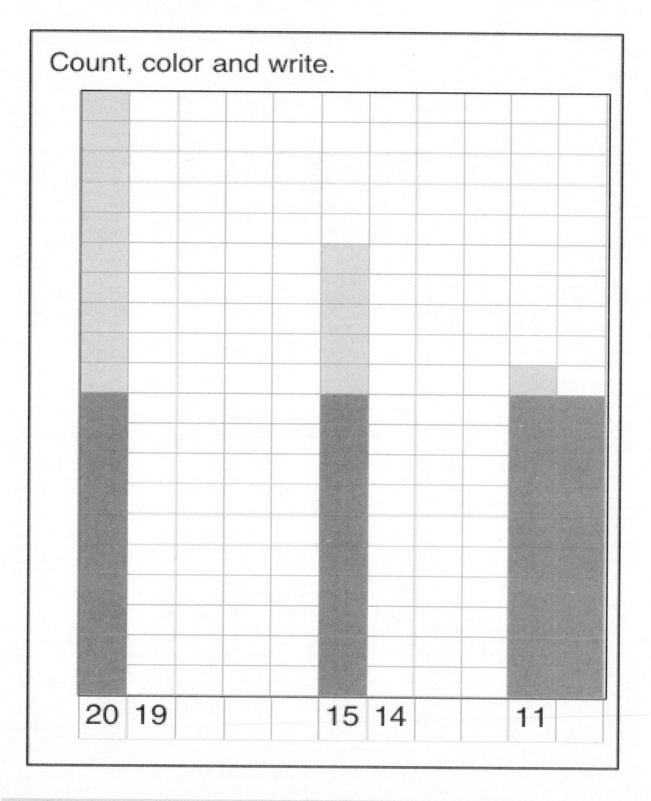

| 20 | 19 | | | | 15 | 14 | | | 11 | |

EUREKA
MATH™

Name _____   Date _____

Write the missing numbers, counting down.

14,	13,	12,	11,	_____
15,	14,	_____,	12,	_____, _____
13,	12,	_____,	_____,	_____

EUREKA MATH™

Lesson 12:   Represent numbers 20 to 11 in tower configurations decreasing by 1—a pattern of *1 smaller*.

©2018 Great Minds®. eureka-math.org

77

Vincent's father made 15 tacos for the family.

Show the 15 tacos as 10 tacos and 5 tacos.

Draw a number bond to match.

 Draw

EUREKA MATH™

Lesson 13:   Show, count, and write to answer *how many* questions in linear and array configurations.

©2018 Great Minds®. eureka-math.org

79

Name _____   Date _____

The ducks found some tasty fish to eat in the boxes!
Count up on the number path.

Write the missing numbers for the boxes that have a duck on top.

_____   _____   _____   _____   _____

Write the missing numbers for the boxes that have a duck on top.

_____   _____   _____   _____   _____

How many ducks do you count?

_____   _____

In the space below, draw 15 circles in rows.

In the space below, draw 12 squares in rows.

Lesson 13:   Show, count, and write to answer *how many* questions in linear and
array configurations.

EUREKA
MATH™

Name _____   Date _____

Count and write how many.

☆ ☆ ☆ ☆
☆ ☆ ☆ ☆
☆ ☆ ☆ ☆   _____

Look at the 3 sets of blocks below. Count the shaded blocks in each set.
Circle the set that has the same number of shaded blocks as stars.

Early finishers: Which was easier to count, stars or blocks? Why?

Lesson 13:   Show, count, and write to answer *how many* questions in linear and
               array configurations.

©2018 Great Minds®. eureka-math.org

83

Eva put 12 cookies on a cookie sheet in 2 rows of 6.

Draw Eva's cookies. Show her 12 cookies as a number bond of 10 ones and

2 ones. Circle the 10 cookies that are inside the 12 cookies.

 Draw

(Give students Hide Zero Cards.) Show the 12 cookies using your Hide Zero cards. Explain to a partner how the parts of the number bond match the parts of your drawing and the Hide Zero cards.

 EUREKA MATH™

Lesson 14:   Show, count, and write to answer *how many* questions with up to 20 objects in circular configurations.

85

©2018 Great Minds®. eureka-math.org

Name _____   Date _____

Whisper count how many objects there are. Write the number.

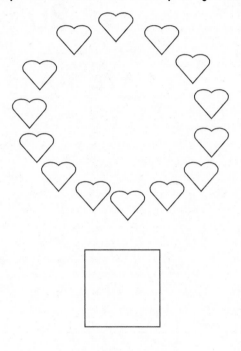

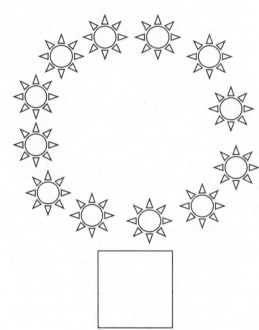

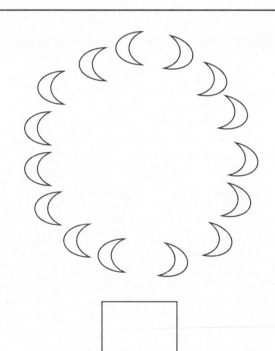

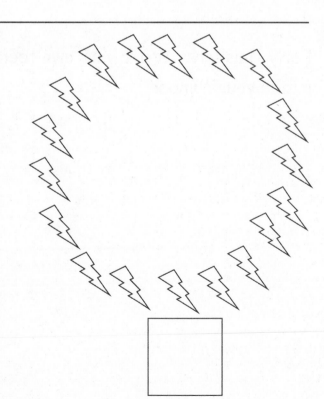

EUREKA MATH™   **Lesson 14:**   Show, count, and write to answer *how many* questions with up to 20 objects in circular configurations.   **87**

©2018 Great Minds®. eureka-math.org

Whisper count and draw in more shapes to match the number.

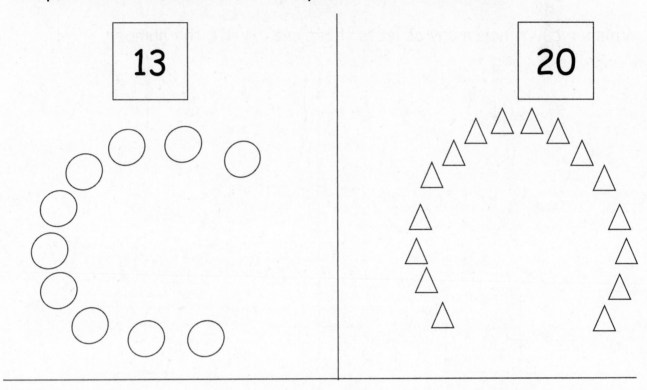

Early finishers: Write your own teen number in the box. Draw a picture to match your number.

EUREKA
MATH™

Name _____   Date _____

Count the stars. Write the number in the box.

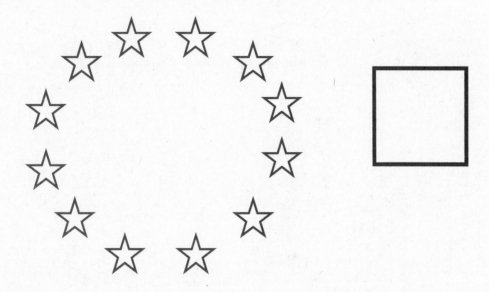

Whisper count and draw in more dots to match the number.

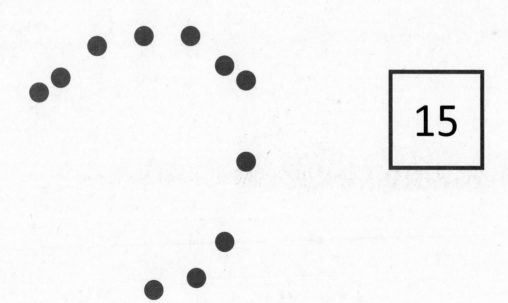

15

EUREKA MATH™   Lesson 14:   Show, count, and write to answer *how many* questions with up to 20 objects in circular configurations.   89

©2018 Great Minds®. eureka-math.org

Mr. Perry is decorating donuts. He puts 14 little dots of chocolate in rows.

Show him an idea about how to put the 14 dots in a circle on his donut.

Draw

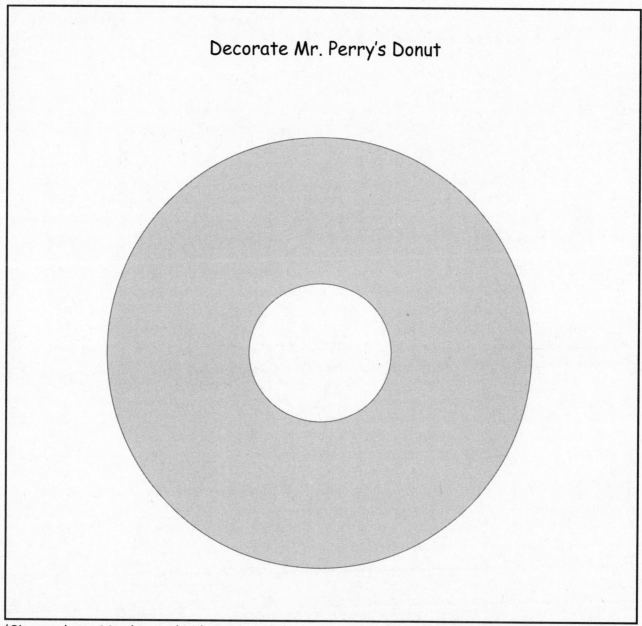

Decorate Mr. Perry's Donut

(Give students 14 cubes and Hide Zero cards.) Use the cubes first, and then draw the chocolate dots on his donut. Show the total number of dots of chocolate with a number bond and the Hide Zero cards.

Name _____   Date _____

Count up by tens, and write the numbers.

	10
	20
	50

Help the puppy down the stairs! Count down by tens. Write the numbers.

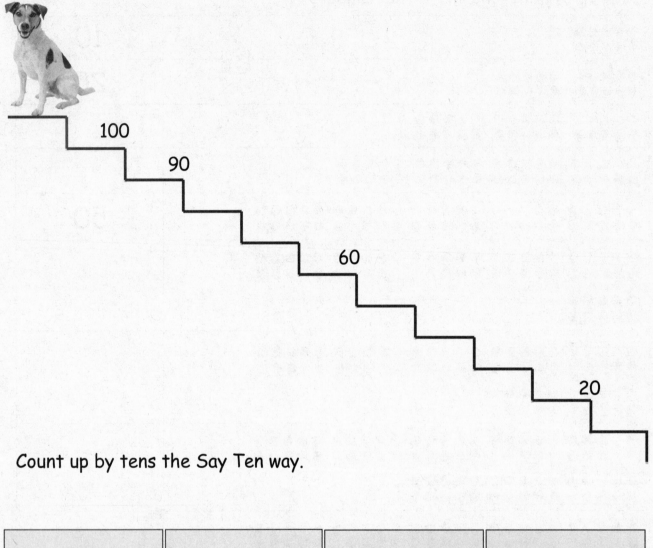

Count up by tens the Say Ten way.

ten	___ tens	_3_ tens	___ tens

___ tens	___ tens	___ ___	___ ___

       Lesson 15:   Count up and down by tens to 100 with Say Ten and regular counting.

EUREKA MATH™

Name _____   Date _____

Count up and down by 10. Write the numbers.

▦	10
▦ ▦	
▦ ▦ ▦	
▦ ▦ ▦ ▦	
▦ ▦ ▦ ▦ ▦	
▦ ▦ ▦ ▦	40
▦ ▦ ▦	
▦ ▦	
▦	

Count down and up by 10 the Say Ten way.

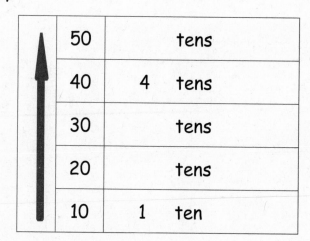

↓	100	10	tens
	90		tens
	80		tens
	70	7	tens
	60		tens

↑	50		tens
	40	4	tens
	30		tens
	20		tens
	10	1	ten

7 students are putting their handprints on a poster board.

How many fingers will show on the poster?

Use the 2-hand cards to help you.

(Give students 2-hand cards.)

Name _____   Date _____

Count up or down by 1s. Help the animals and the girl get what they want!

| 20 | | 22 | | 24 | | 26 | | | |

40 · · · 44 · 46 · 48

92 · · · · · · 98 · 99

Count up.

| 63 | 64 | | | |

Stop!

Count down.

| 66 | | | |

Name _____   Date _____

Help the cow get to the barn by counting by 1s.

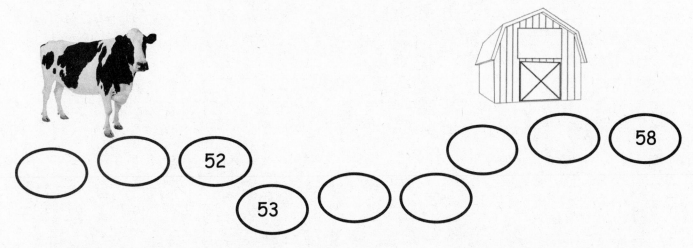

Help the boy get to his present. Count up by 1s. When you get to the top, count down by 1s.

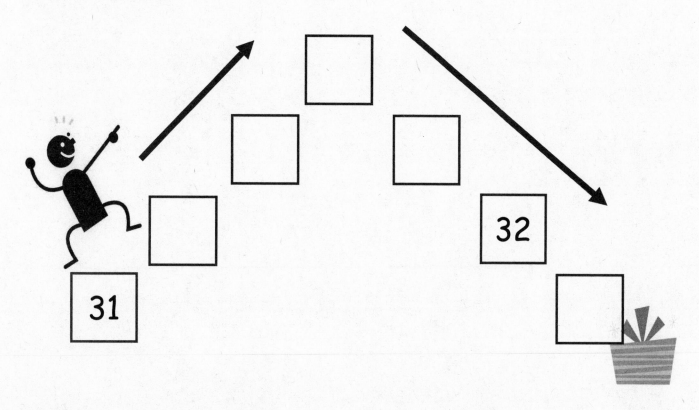

Sammy's mom has 10 apples in a bag.

Some are red and some are green.

What might be the number of each color apple in her bag?

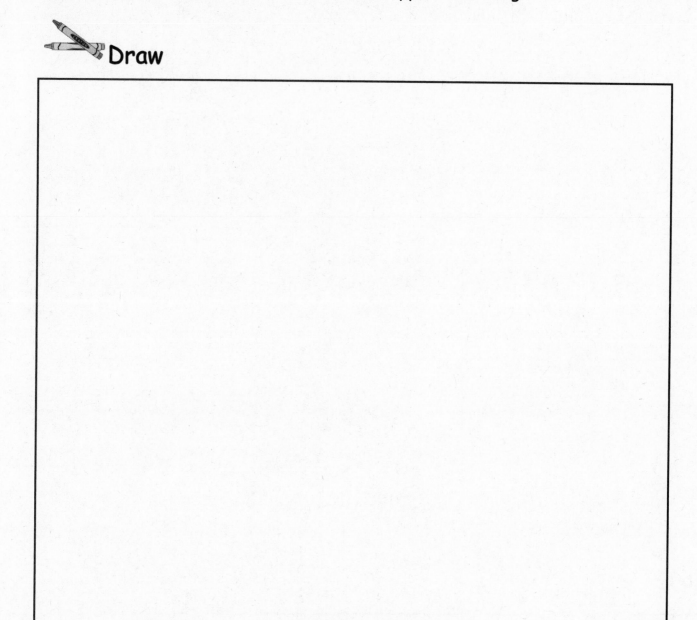

Draw

There is more than one possible answer. See how many different answers can be found. Show the answers with number bonds. Label the parts as R and G.

Lesson 17:   Count across tens when counting by ones through 40.

©2018 Great Minds®. eureka-math.org

103

Name _____   Date _____

Touch and count the dots from left to right starting at the arrow. Count to the puppy, and then keep counting to his bones and twin brother!

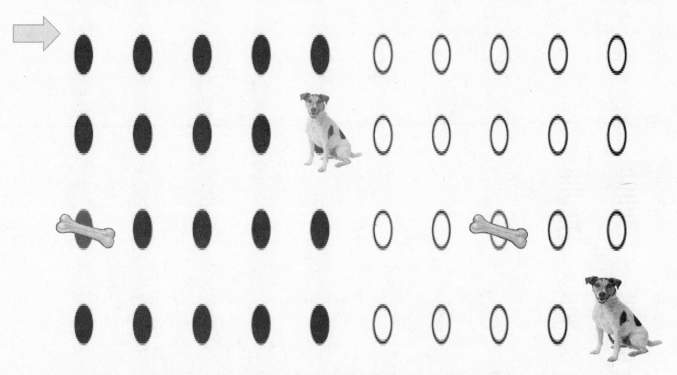

Count again and color the last dot of each row green. When you have finished, go back and see if you can remember your green numbers!

What number did you say when you touched the first puppy?

- The first bone?

- The second bone?

- His twin brother?

Count each number by 1s. Write the number below when there is a box.

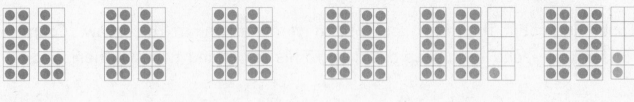

| 17 | | | | 21 | |

Touch and count the rocks from the cow to the grass!

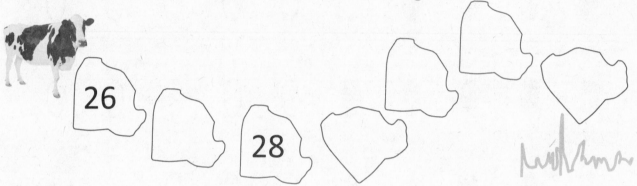

26   28

Count up by 1s. Help the kitty play with her yarn!

| 31 | 32 | | | | 36 | | | | |

Count down by 1s.

11   10   21   19   31

EUREKA MATH

Name _____   Date _____

Touch and count carefully. Cross out the mistake, and write the correct number.

Example:

3

1, 2, ~~9~~, 4, 5

20	21	22	23	24	25	29

30	31	32	33	43	35	36

25	26	27	28	29	29	31

34	35	36	37	38	39	44

EUREKA
MATH™

Susan puts 9 flowers in two vases.

Draw the flowers to show a way she might do that.

Make a number bond and a number sentence to match your picture.

 Draw

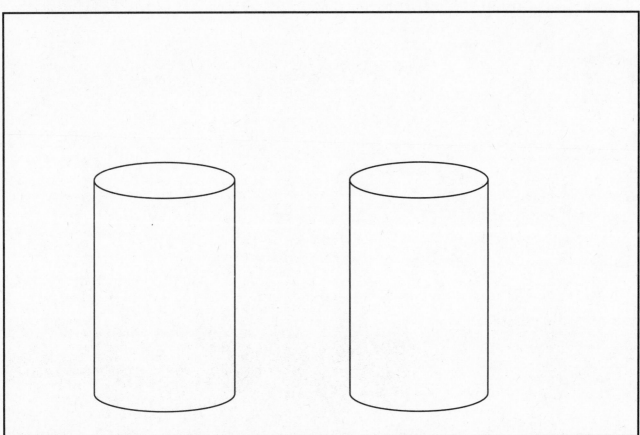

Extension: See if there is another way to put the flowers in the vases.

 Write

Teachers' Directions for the Rekenrek Problem Set

Have students show 50 dots by using their hiding paper to cover the other rows.

Then, have students whisper count all the dots. Say the last number in each row loudly, and color the circle green.

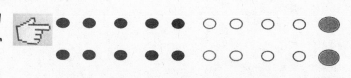
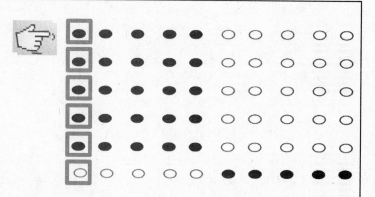

Have students show 60 dots using their hiding paper to cover 4 rows.

Then, have students whisper count all the dots. Have them box the first dot in each row with blue and say its number loudly.

Have students show 70 dots by hiding 30 dots.

Then, have students whisper count all the dots. Have them put a triangle around the fifth dot in each row with red and say those numbers loudly.

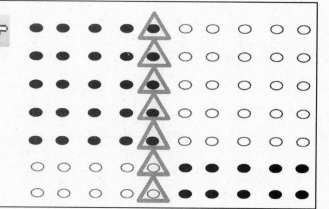

EUREKA MATH™

Lesson 18:   Count across tens by ones to 100 with and without objects.

111

©2018 Great Minds®. eureka-math.org

Name _____   Date _____

Rekenrek

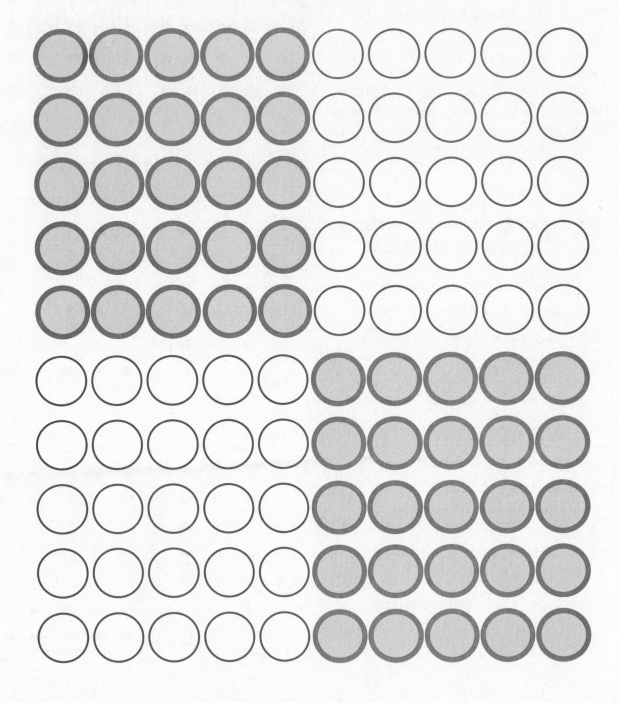

EUREKA
MATH™

Name _____   Date _____

Touch and whisper count the circles by 1s to 100. Say the last number in each row loudly, and color the circle purple. Do your best. Your teacher may call time before you are finished.

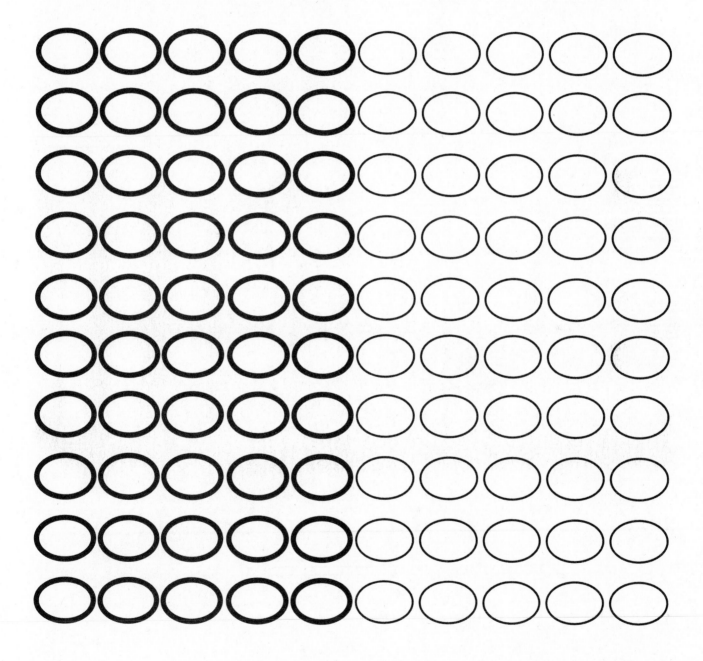

EUREKA MATH™

©2018 Great Minds®. eureka-math.org

The light is out, and it's dark.

Peter left 7 blue and green beads for his crafts on his desk.

But he can't see how many are blue or how many are green in the dark.

Draw a picture to show what the colors of his beads might be when he turns on the light.

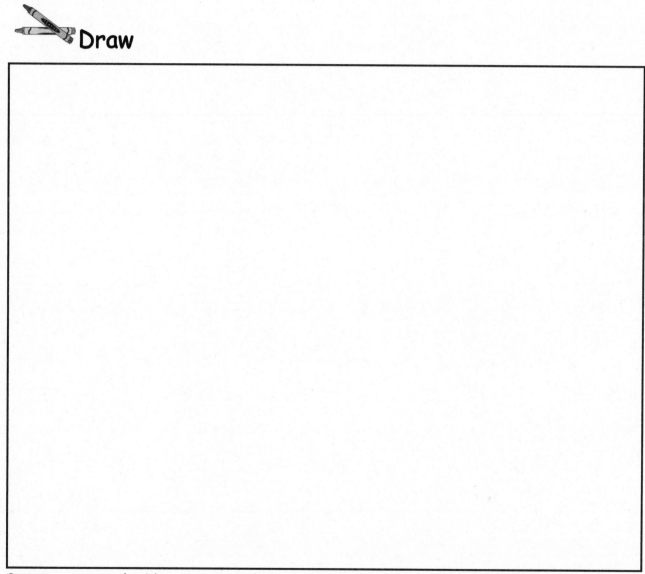**Draw**

Compare your work with a partner. Did you show your beads the same way? Why or why not?

How is this problem like the problems in previous lessons with the flowers and the apples?

 Lesson 19: Explore numbers on the Rekenrek. (Optional)

115

©2018 Great Minds®. eureka-math.org

Name _____   Date _____

Find the Hidden Teen Number

Show each number on your Rekenrek with your partner. Write how many.
Circle the teen number inside the big number. Draw a line from the big
number to the teen number that hides inside it.

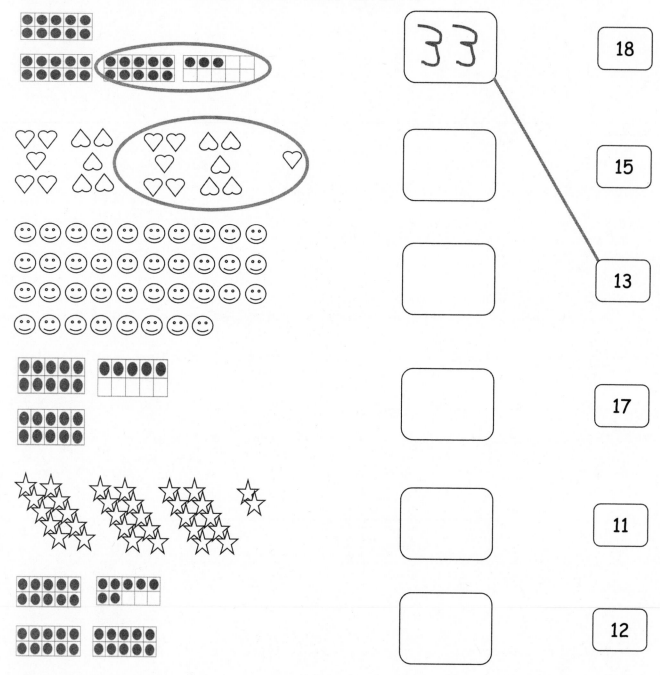

EUREKA
MATH™

Lesson 19:   Explore numbers on the Rekenrek. (Optional)

117

©2018 Great Minds®. eureka-math.org

Name _____   Date _____

Show the number on your Rekenrek with your partner. In the box, write the number that tells how many objects there are. Circle the teen number you see. Write the teen number in the other box.

EUREKA
MATH™

Lesson 19:   Explore numbers on the Rekenrek. (Optional)

©2018 Great Minds®. eureka-math.org

119

Each student was given 6 colored pencils and 4 regular pencils.

How many pencils did each student get?

Draw a picture and a number bond. Then, write a number sentence.

 Draw

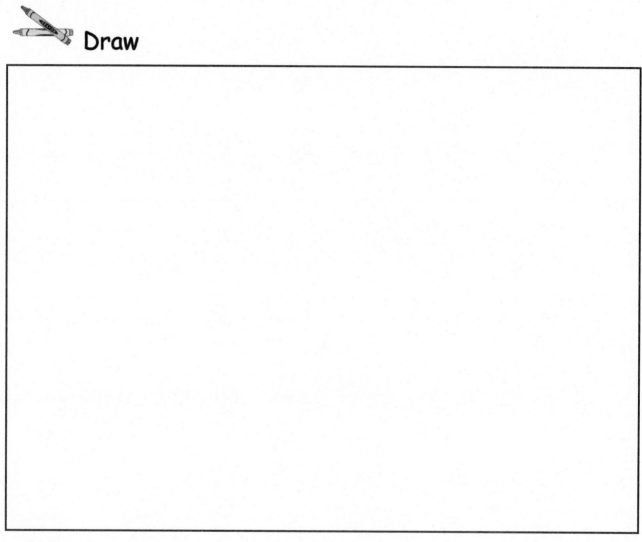

Write

Lesson 20:   Represent teen number compositions and decompositions as addition
                sentences.

©2018 Great Minds®. eureka-math.org

121

Name _____   Date _____

Fill in each number bond, and write a number sentence to match.

Example:

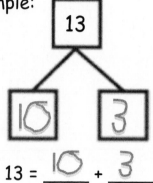

13 = __10__ + __3__

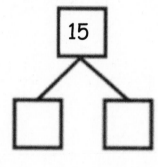

15 = _____ + _____

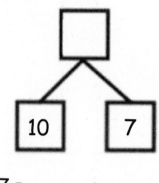

17 = _____ + _____

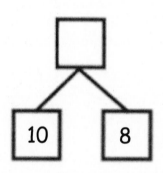

10 + 8 = _____

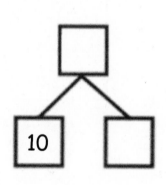

10 + 6 = _____

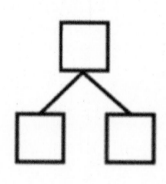

_____ = 10 + 4

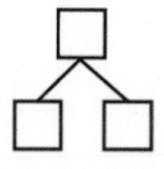

12 = _____ + _____

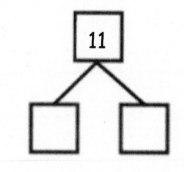

_____ = _____ + _____

Early finishers: Make up your own teen number bonds and number sentences on the back!

Name _____   Date _____

The first number is the whole. Circle its parts. | 5 | 1 | ②| ③ |

| 12 | 10 | 6 | 2 |

| 11 | 1 | 10 | 8 |

| 14 | 4 | 2 | 10 |

| 18 | 1 | 10 | 8 |

| 10 | 10 | 1 | 0 |

| 20 | 10 | 2 | 10 |

EUREKA MATH™

Lesson 20:   Represent teen number compositions and decompositions as addition sentences.

©2018 Great Minds®. eureka-math.org

125

Peter saw 8 puppies at the pet store.

While he was watching them, 2 hid in a box.

How many puppies could Peter see then?

Draw a picture, and write a number bond and number sentence to match the story.

 Draw

Lesson 21:   Represent teen number decompositions as 10 ones and some ones, and find a hidden part.

©2018 Great Minds®. eureka-math.org

127

Lesson 21:   Represent teen number decompositions as 10 ones and some ones, and find a hidden part.

EUREKA
MATH™

Name _____   Date _____

Model each number with cubes on your number bond mat. Then, complete the number sentences and number bonds.

Example:
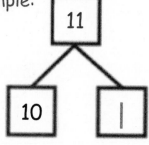   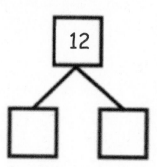   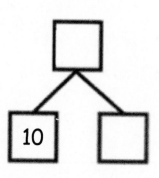

11 = 10 + ___|___   12 = 10 + _____   13 = 10 + _____

10 + __|__ = 11   10 + _____ = 12   10 + _____ = 13

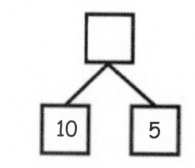   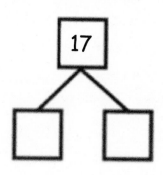   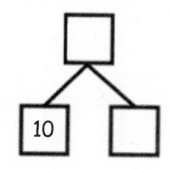

_____ + 5 = 15   _____ + 7 = 17   _____ + 8 = 18

15 = _____ + 5   17 = _____ + 7   18 = 10 + _____

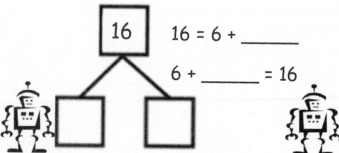

16 = 6 + _____

6 + _____ = 16

9 + _____ = 19

19 = 10 + _____

Name _____   Date _____

Complete the number sentences and number bonds. Use your materials to help you.

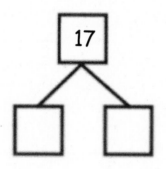

_____ + 7 = 17   17 = _____ + 10

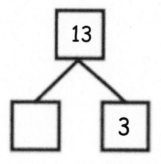

_____ + 3 = _____   13 = _____ + 10

EUREKA MATH™   **Lesson 21:**   Represent teen number decompositions as 10 ones and some ones,   131
                                   and find a hidden part.

©2018 Great Minds®. eureka-math.org

Lisa has 5 pennies in her hand and 2 in her pocket.

Draw Lisa's pennies.

Matt has 6 pennies in his hand and 2 in his pocket.

Draw Matt's pennies.

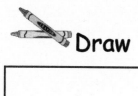Draw

Who has fewer pennies? How do you know?

Lesson 22:   Decompose teen numbers as 10 ones and some ones; Compare *some ones* to compare the teen numbers.

©2018 Great Minds®. eureka-math.org

133

Name _____   Date _____

Circle 10 erasers. Circle 10 pencils. Match the extra ones to see which group has more. ✓ Check the group that has *more* things.

Circle 10 sandwiches. Circle 10 milk cartons. ✓ Check the group that has *less* things.

Circle 10 baseballs. Circle 10 gloves. Write how many are in each group. ✓ Check the group that has *more* things.

EUREKA MATH™   Lesson 22:   Decompose teen numbers as 10 ones and some ones; compare *some ones* to compare the teen numbers.   135

©2018 Great Minds®. eureka-math.org

Circle 10 apples. Circle 10 oranges. Write how many are in each group.
✓ Check the group that has *less*.

 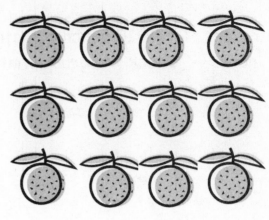

Circle 10 spoons. Circle 10 forks. Write how many are in each group.
Circle *more* or *less*.

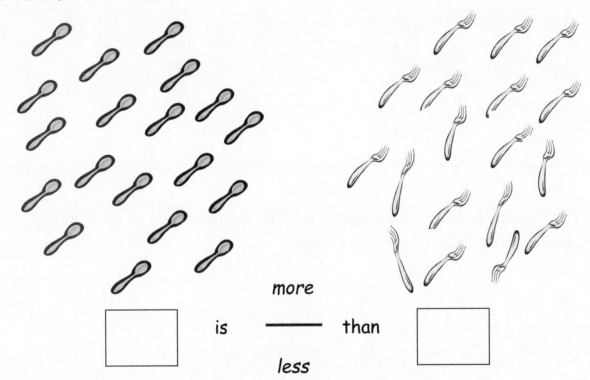

more
is ———— than
less

Lesson 22:   Decompose teen numbers as 10 ones and some ones; compare *some*
                                   ones to compare the teen numbers.

                    ©2018 Great Minds®. eureka-math.org

EUREKA MATH™

Name _____   Date _____

Count and write the number.
Circle *more* or *less*.

1   is   (less) $\underset{\text{\_\_\_\_}}{\overset{\text{more}}{}}$ than 4

$\underline{\quad}$ is $\overset{\text{more}}{\underline{\text{less}}}$ than $\underline{\quad}$

$\underline{\quad}$ is $\overset{\text{more}}{\underline{\text{less}}}$ than $\underline{\quad}$

$\underline{\quad}$ is $\overset{\text{more}}{\underline{\text{less}}}$ than $\underline{\quad}$

Lesson 22:   Decompose teen numbers as 10 ones and some ones; compare *some ones* to compare the teen numbers.

137

©2018 Great Minds®. eureka-math.org

Name _____   Date _____

Robin sees 5 apples in a bag and 10 apples in a bowl. Draw a picture to show how many apples there are.

Write a number bond and an addition sentence to match your picture.

_____ _____ _____

Sam has 13 toy trucks. Draw and show the trucks as 10 ones and some ones.

Write a number bond and an addition sentence to match your picture.

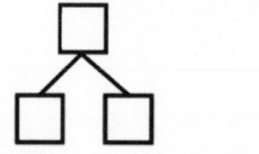

_____ _____ _____

Lesson 23:   Reason about and represent situations, decomposing teen numbers into 10 ones and some ones and composing 10 ones and some ones into a teen number.

EUREKA MATH

©2018 Great Minds®. eureka-math.org

139

Our class has 16 bags of popcorn. Draw and show the popcorn bags as 10 ones and some ones.

Write a number bond and an addition sentence to match your picture.

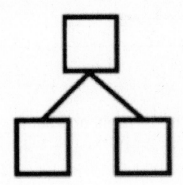

_____ _____ _____

Lesson 23:   Reason about and represent situations, decomposing teen numbers into 10 ones and some ones and composing 10 ones and some ones into a teen number.

©2018 Great Minds®. eureka-math.org

EUREKA
MATH™

Name _____   Date _____

There are 12 balls. Draw and show the balls as 10 ones and **some ones**.

Write a number bond to match your picture.

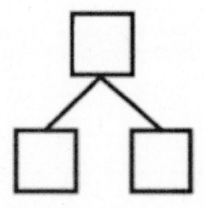

Write an addition sentence to match your number bond.

_____ _____ _____

picture and word problem

Lesson 23: Reason about and represent situations, decomposing teen numbers into 10 ones and some ones and composing 10 ones and some ones into a teen number.

143

©2018 Great Minds®. eureka-math.org

Grade K
Module 6

Draw several things you saw this past week that looked like shapes
you know.

Draw

Share your picture with your partner. What are the different shapes called? Talk about each of the
shapes and how you knew its name. Does your partner agree with you?

Name _____   Date _____

Listen to the directions.

First, draw the missing line to finish the triangle using a ruler. **Second**, color the corners red. **Third**, draw another triangle.

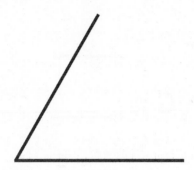

First, use your ruler to draw 2 lines to make a square. **Second**, color the corners red. **Third**, draw another square.

First, draw a triangle using your ruler. **Second,** draw a different triangle using your ruler. **Third**, show your pictures to your partner.

Lesson 1:   Describe the systematic construction of flat shapes using ordinal numbers.

©2018 Great Minds®. eureka-math.org

149

$4 + 1 =$ _____

_____ $= 2 + 1$

$3 + 2 =$ _____

$3 + 1 =$ _____

_____ $= 5 + 0$

$5 - 1 =$ _____

_____ $= 4 - 1$

$3 - 2 =$ _____

$3 - 0 =$ _____

_____ $= 5 - 4$

$2 - 1 =$ _____

_____ $= 3 - 3$

$1 - 0 =$ _____

$3 - 0 =$ _____

_____ $= 4 - 4$

$2 + 2 =$ _____

_____ $= 5 - 3$

$1 + 1 =$ _____

$4 - 0 =$ _____

_____ $= 4 + 1$

Lesson 1:   Describe the systematic construction of flat shapes using ordinal numbers.

EUREKA
MATH™

Name _____   Date _____

Use your ruler.

First, draw a straight line from the dot.

Second, draw a different straight line from the dot.

Third, draw another straight line to make a triangle.

●

Lesson 1:   Describe the systematic construction of flat shapes using ordinal
             numbers.

©2018 Great Minds®. eureka-math.org

151

Name _____   Date _____

First, use a ruler to trace the shapes. Second, follow the directions in each box. Use your ruler to draw the shapes.

Draw 3 different triangles.

Draw 2 different rectangles.

Draw 1 hexagon.

Lesson 2:   Build flat shapes with varying side lengths and record with drawings.

153

©2018 Great Minds®. eureka-math.org

5 - 4 = _____

5 - 3 = _____

5 - 2 = _____

5 - 1 = _____

5 - 0 = _____

0 + 1 = _____

1 + 1 = _____

2 + 1 = _____

3 + 1 = _____

4 + 1 = _____

4 - 2 = _____

2 - 1 = _____

3 - 2 = _____

3 - 1 = _____

5 - 0 = _____

4 - 3 = _____

2 + 1 = _____

3 + 2 = _____

4 - 1 = _____

5 - 4 = _____

Lesson 2:   Build flat shapes with varying side lengths and record with drawings.

EUREKA MATH™

Name _____   Date _____

First, draw a triangle so all of the sides are different lengths.

Second, draw a triangle with your ruler that has 2 sides that are about the same length.

Connect the dots to make a shape with 3 sides.

Connect the dots to make a shape with 4 sides.

Draw

(Instruct students to use a straight edge to connect the dots to make the shapes.) Compare your 3-sided shape to your friend's. Do they look the same? Name the shapes. Compare your 4-sided shape to your friend's. Do they look the same? Name the shapes.

Lesson 3:   Compose solids using flat shapes as a foundation.

157

©2018 Great Minds®. eureka-math.org

Name _____ Date _____

Trace the circles and rectangle. Cut out the shape. Fold and tape to create a cylinder.

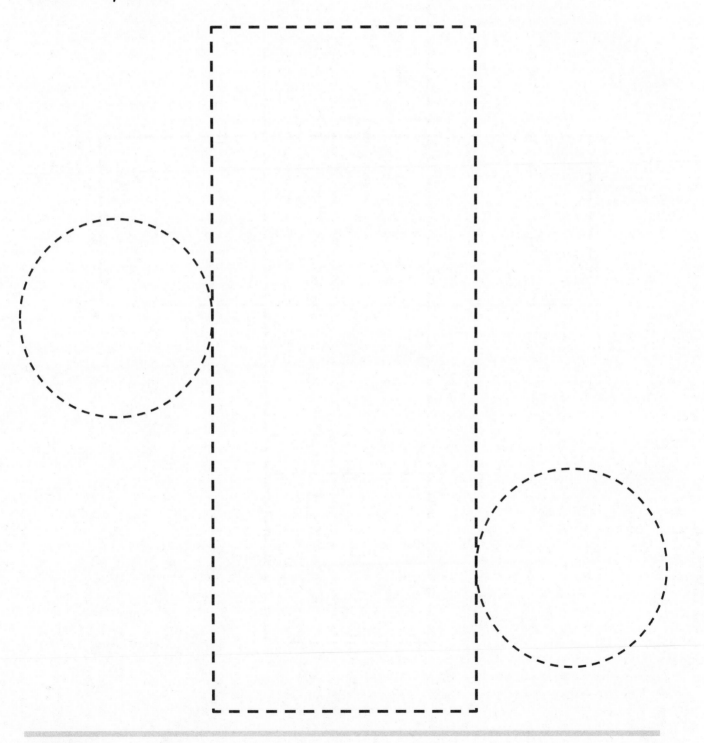

Trace the squares. Cut out the shape. Fold and tape to create a cube.

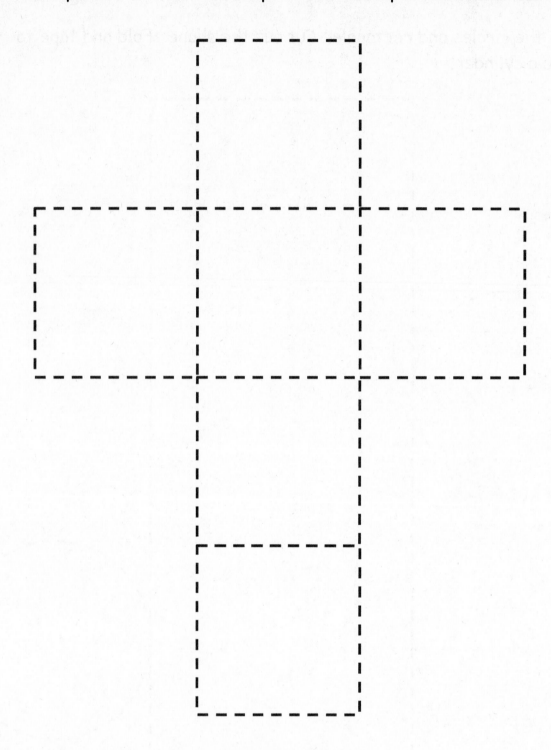

Lesson 3:   Compose solids using flat shapes as a foundation.

Name _____   Date _____

Draw a line from the flat shape to the object that has a face with that flat shape.

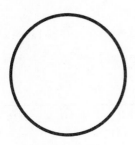

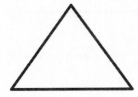

Draw 3 three-sided shapes.

Draw 4 four-sided shapes.

Complete the number bond, and write a number sentence to tell how many shapes you drew in all.

 Draw

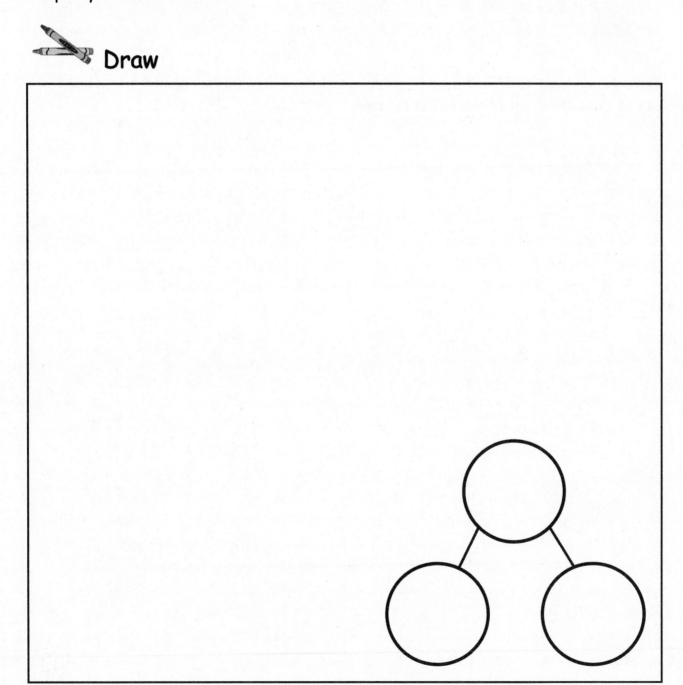

Write

Compare your work to a friend's. Do your shapes look the same? Do your number bonds look the same? Do your number sentences look the same?

Lesson 4:   Describe the relative position of shapes using ordinal numbers.

Name _____   Date _____

Circle the 2ⁿᵈ truck from the stop sign. Draw a square around the
5ᵗʰ truck. Draw an X on the 9ᵗʰ truck.

Draw a triangle around the 4ᵗʰ vehicle from the stop sign. Draw a circle
around the 1ˢᵗ vehicle. Draw a square around the 6ᵗʰ vehicle.

Put an X on the 10ᵗʰ horse from the stop sign. Draw a triangle around the
7ᵗʰ horse. Draw a circle around the 3ʳᵈ horse. Draw a square around the
8ᵗʰ horse.

**EUREKA
MATH**™

Lesson 4:   Describe the relative position of shapes using ordinal numbers.

©2018 Great Minds®. eureka-math.org

165

Draw a line from the shape to the correct ordinal number, starting at the top.

9th ninth	
4th fourth	
6th sixth	
1st first	
7th seventh	
3rd third	
10th tenth	
5th fifth	
8th eighth	
2nd second	

Lesson 4:   Describe the relative position of shapes using ordinal numbers.

EUREKA MATH™

Name _____   Date _____

Listen to the directions. Start at the circle when counting.

Color the 5th shape red.

Color the 2nd shape green.

Color the 10th shape yellow.

Color the 7th shape blue.

Color the 1st shape pink.

Color the 8th shape orange.

_____

shapes

Listen as we draw a house together.

 Draw

First, draw a square for the big part of your house. Second, make a triangle roof. Third, make a door of any shape. Fourth, find somewhere in your picture to use two more squares or rectangles. Fifth, use a circle in your picture. Sixth, use a hexagon in your picture. Don't forget to draw yourself. Show your picture to your partner. Do your houses look alike? How did you use the shapes differently in your pictures?

EUREKA
MATH™

Lesson 5:   Compose flat shapes using pattern blocks and drawings.

171

©2018 Great Minds®. eureka-math.org

Name _____   Date _____

Choose 4 shapes to create a new shape in Box 1. Give the same 4 shapes to your partner. Have your partner create a different shape in Box 2.

1

2

Lesson 5:   Compose flat shapes using pattern blocks and drawings.

173

©2018 Great Minds®. eureka-math.org

Choose 5 shapes to create a new shape in Box 3. Give the same 5 shapes to your partner. Have your partner create a different shape in Box 4.

3

4

Subtract.

$5 - 1 = \boxed{}$ $5 - 2 = \boxed{}$ $5 - 3 = \boxed{}$ $5 - 4 = \boxed{}$

EUREKA MATH

Name _____   Date _____

Use your pattern blocks to help you solve the problem.

Use 2 blocks to make a rectangle. Trace your blocks to show your rectangle.

I Can Make New Shapes!

_____

I can make new shapes recording sheet

Lesson 5:   Compose flat shapes using pattern blocks and drawings.

177

©2018 Great Minds®. eureka-math.org

Look around the classroom to find something made of more than one shape.

Draw what you found.

Use a marker to trace the shapes in the object you drew.

Draw

(If necessary, give hints about items with more than one shape that can be drawn. Help students find and highlight the shapes within shapes in their drawings.)

Name _____   Date _____

Trace to show 2 ways to make each shape. How many shapes did you use?

I used __3__ shapes.

I used _____ shapes.

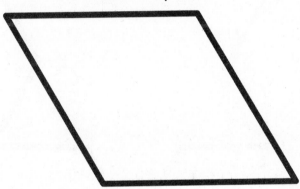

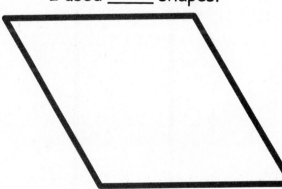

I used _____ shapes.

I used _____ shapes.

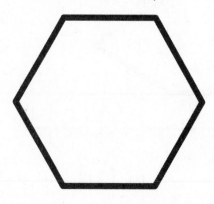

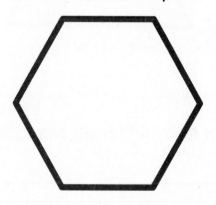

I used _____ shapes.

I used _____ shapes.

EUREKA MATH™

Lesson 6:   Decompose flat shapes into two or more shapes.

181

©2018 Great Minds®. eureka-math.org

Fill in each shape with pattern blocks. Trace to show the shapes you used.

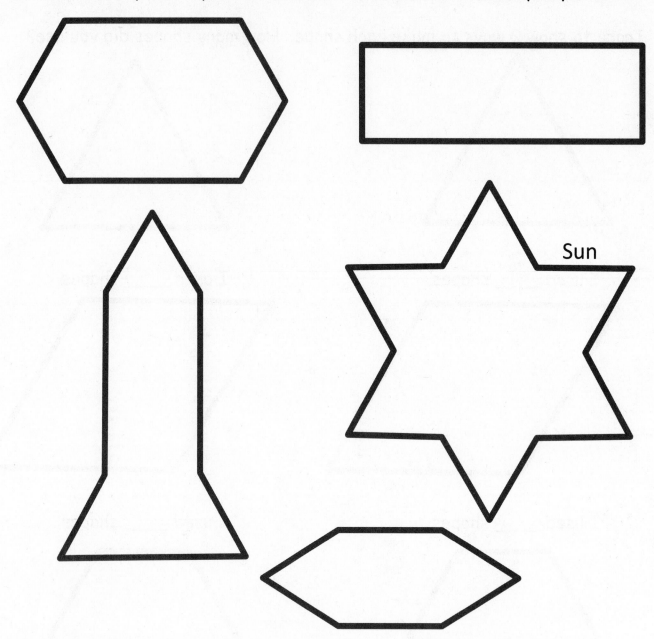

Sun

How many different ways can you cover the sun picture with pattern blocks?

EUREKA
MATH™

Name _____   Date _____

Draw 2 shapes that can be used to build the rectangle.

Draw 2 shapes that can be used to build the house.

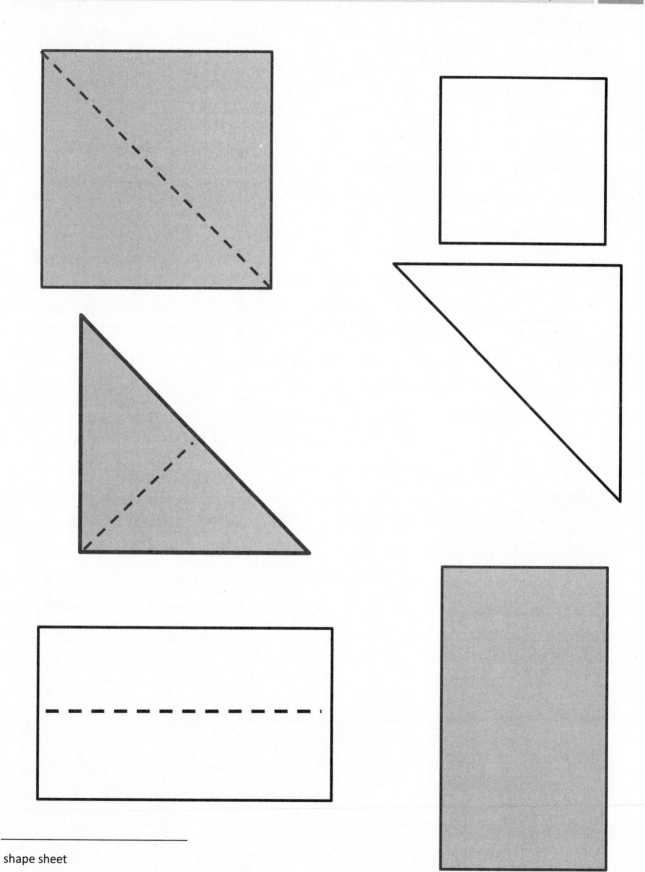

shape sheet

Draw a big rectangle and pretend it is a cake.

Draw 2 lines to cut the cake.

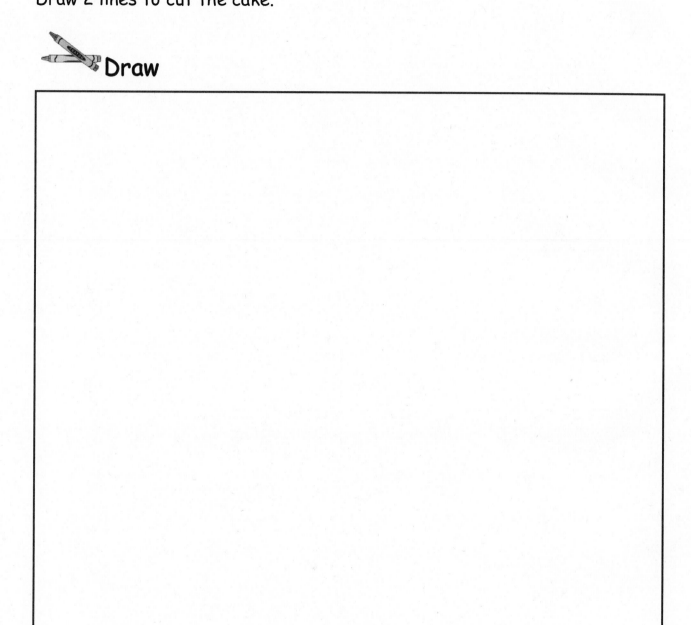 Draw

Look at how your neighbor cut his cake. Did you both cut the cake the same way? Do you both have the same number of pieces? Does one of you have more pieces? Does one of you have fewer pieces?

Name _____   Date _____

Glue your puzzles into the frames.

Glue puzzle here.

Glue puzzle here.

Draw some of the shapes that you had after you cut your rectangles.

EUREKA MATH™

Lesson 7:   Compose simple shapes to form a larger shape described by an outline.

©2018 Great Minds®. eureka-math.org

189

Carlos drew 2 lines on his square. You can see his square before he cut it.
Circle the shapes Carlos had after he cut.

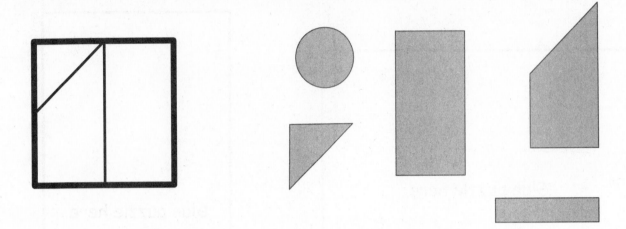

India drew 2 lines on her rectangle. You can see her rectangle before she
cut it. Circle the shapes India had after she cut.

Lesson 7:   Compose simple shapes to form a larger shape described by an outline.

EUREKA
MATH™

Name _____   Date _____

If you drew 2 straight lines inside the gray rectangle, what shapes might you find? Circle them.

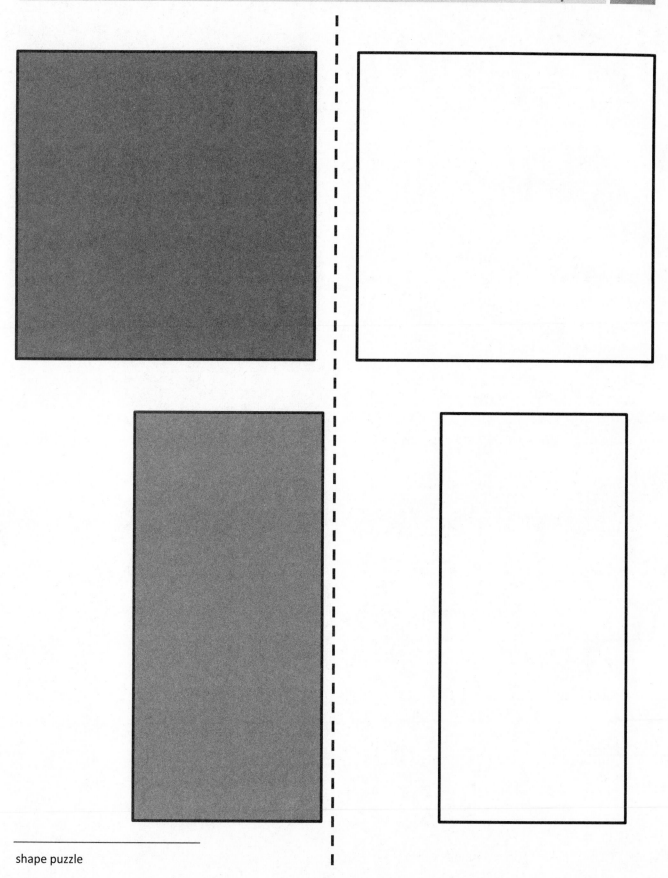

shape puzzle

EUREKA
MATH™

Name _____   Date _____

A. Make-10 Mania: Show how you made 10.

- -

Name _____   Date _____

B. Five-Group Frenzy: Write the number, draw the number in the
5-group way, and draw the number in any other configuration.

Lesson 8:   Culminating task—review selected topics to create a cumulative year-end   **195**
                project.

©2018 Great Minds®. eureka-math.org

Name _____ Date _____

C. Shape Shifters: Choose 5 pattern blocks, and create a shape. Trace your shape, and then trade with a partner.

- -

Name _____ Date _____

D. The Weigh Station: Choose an object. Guess how many pennies are the same weight as the object. Then, see if you guessed correctly! Draw a picture of the object, and write how many pennies it weighs.

Lesson 8:   Culminating task—review selected topics to create a cumulative year-end project.

©2018 Great Minds®. eureka-math.org

197

Name _____   Date _____

E. Awesome Authors: Roll the die. Use the number to create an addition or take-away sentence. Draw a picture, number bond, and number sentence. Share your story with a friend.

Lesson 8:   Culminating task—review selected topics to create a cumulative year-end project.

©2018 Great Minds®. eureka-math.org

199

Credits

Great Minds® has made every effort to obtain permission for the reprinting of all copyrighted material. If any owner of copyrighted material is not acknowledged herein, please contact Great Minds for proper acknowledgment in all future editions and reprints of this module.